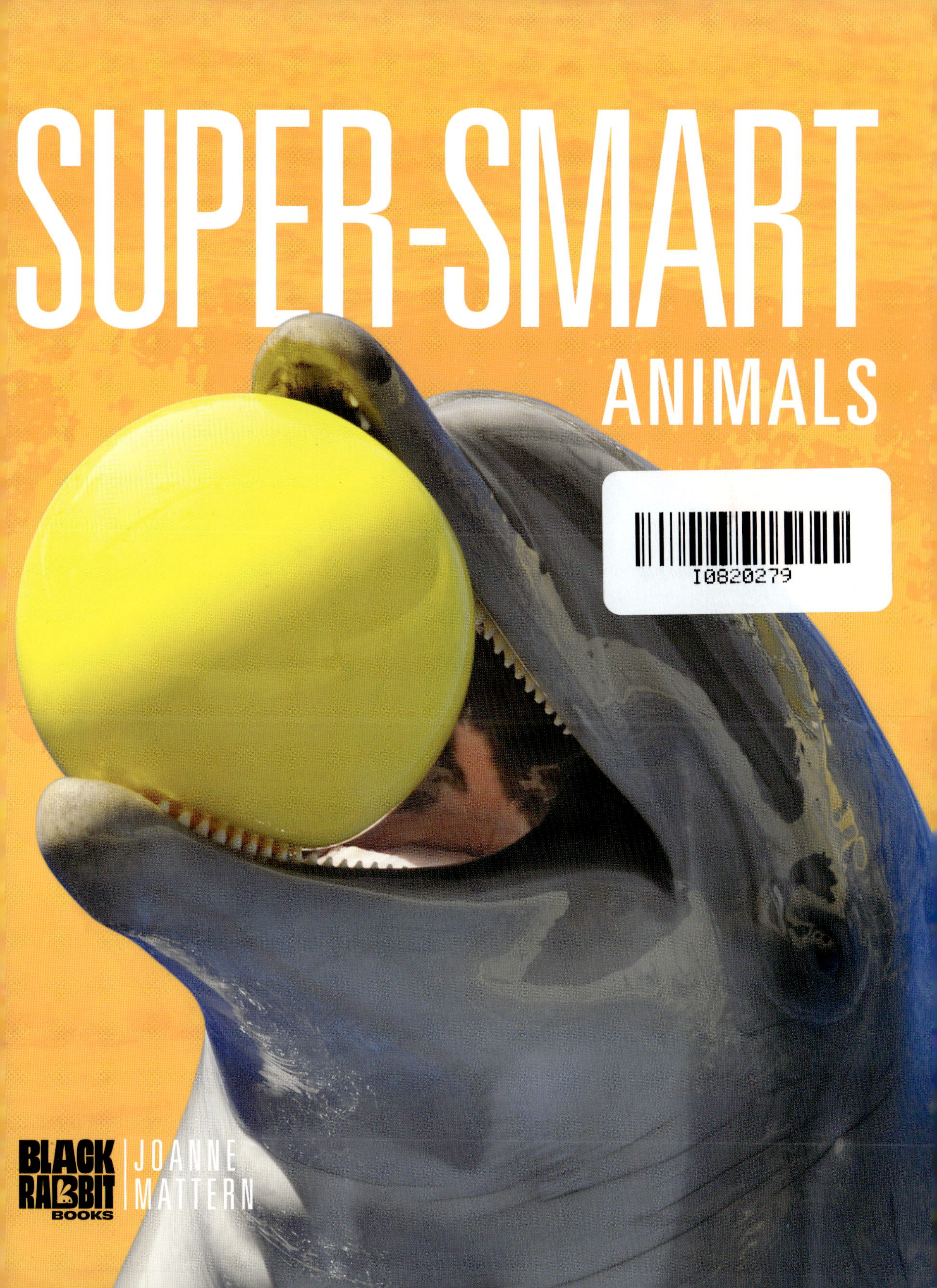
SUPER-SMART
ANIMALS
JOANNE
MATTERN
BLACK
RABBIT
BOOKS

TABLE OF CONTENTS

1

The Octopus

An octopus might not look smart. But this sea creature has the largest brain of any invertebrate. This big brain helps the octopus stay alive. An octopus learns how to hide from predators. It knows how to use tools to make shelters for itself.

Scientists learned that an octopus can solve puzzles and mazes. An octopus is also a great escape artist. They are smart enough to break out of their tanks in aquariums.

Did You Know?
An octopus often plays catch and other games with objects in the water.

2

The Orangutan

7

Orangutans are primates. Apes, monkeys, and people are primates too. Primates have big brains. They are smart creatures.

Orangutans are some of the smartest primates. These animals are great at using tools. They use sticks to fish. They use rocks to crush nuts and bugs.

Young orangutans stay with their mothers for six or seven years. Their mothers teach them how to live. Young orangutans have a lot to learn. It's a good thing they are so smart!

Did You Know?
Orangutans can solve puzzles. They can even use touch screens on computers!

3

The Bottlenose Dolphin

Bottlenose dolphins are smart sea animals. Some scientists think dolphins might be the second smartest animal on Earth. People being the smartest.

Dolphins and people seem very different. But in some ways, they are alike. Dolphins can "talk" to each other. But instead of words, they use whistles and clicking sounds. They also shake their heads and fins to talk to other dolphins.

Dolphins also have super-sized brains. A dolphin's brain can be bigger than a person's.

Think About It

Does it make sense to compare how smart dolphins and people are? Why or why not?

The Rat

4

Scientists spend a lot of time with rats. That is because rats are super-smart!

Rats have a great memory. They can remember faces and smells. They also remember things they have done in the past.

Scientists study rats. They figure out how rats remember patterns. Scientists hope that watching rats can help people. It can help us understand how a person's brain works too.

Think About It

How can studying rats and other animals help people learn about themselves?

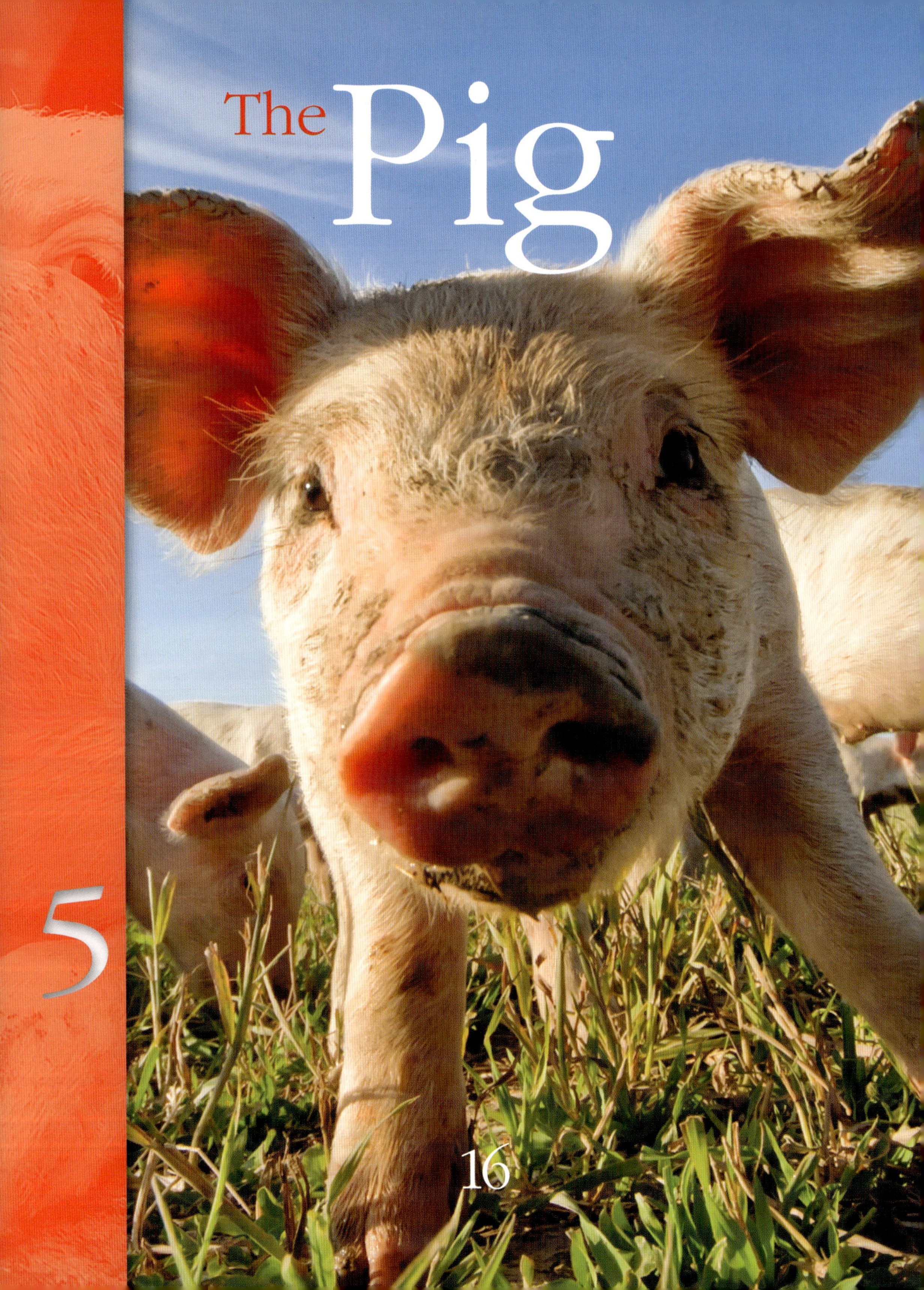
The Pig
5
16

A pig can do a lot more than lay in the mud. Scientists think pigs are the smartest **domestic** animal. They are even smarter than dogs.

Pigs have great memories. They can solve problems. A pig even learns by watching another pig figure out a problem. Pigs also remember people and places. Some scientists also think they can tell time.

Did You Know?

Pigs live in big groups. They care about each other. They help each other. That is another sign that they are smart.

The Crow

An old story describes how a crow gets water by dropping pebbles into a pitcher. This is more than just a story. Crows are smart enough to solve many problems.

Scientists watch crows figure out how to get water out of a tube. The crows drop rocks into the tube to make the water rise. Then the crows surprised the scientists. The birds figured out that dropping stones into one tube would raise the water level in a different tube. This is something even kids ages four to six can't figure out.

Did You Know?
Crows are great at using tools. They bend sticks to catch bugs.

MORE TO EXPLORE

WHERE IN THE WORLD?

PACIFIC OCEAN

NORTH AMERICA

ATLANTIC OCEAN

EUROPE

ASIA

PACIFIC OCEAN

SOUTH AMERICA

AFRICA

INDIAN OCEAN

AUSTRALIA

OCTOPUSES
All oceans

ORANGUTANS
Southeast Asia

BOTTLENOSE DOLPHINS
All oceans

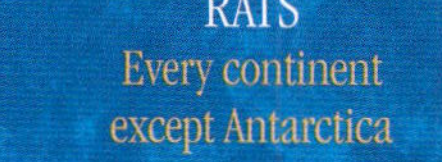

RATS
Every continent except Antarctica

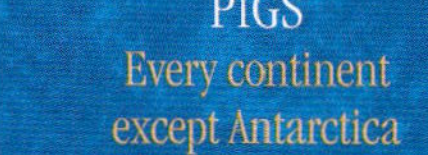

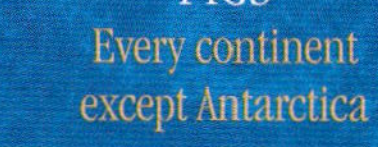

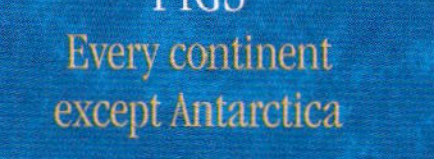

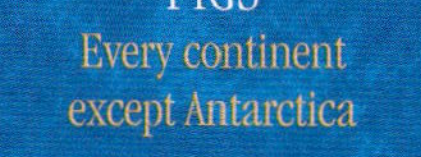

PIGS
Every continent except Antarctica

CROWS
Every continent except South America and Antarctica

MORE TO EXPLORE

FANTASTIC FACTS

An octopus squirts a dark blue liquid to escape from predators.

An orangutan named Rocky learned how to copy human speech.

Dolphins recognize themselves in the mirror. Most other animals do not.

Rats make good pets. They are easy to train and they love to learn new tricks.

Pigs can learn how to play video games.

Crows can remember people's faces. They warn each other when they see someone they don't like.

MORE TO EXPLORE

COOL COMPARISONS

What are some super-smart things the animals in this book have done?

Octopus

Orangutan

Bottlenose Dolphin

Rat

Pig

Crow

• Escape from tanks and cages • Solve puzzles • Solve problems • Use tools • Play catch • Learn patterns • Care for each other • Communicate • Have good memories

MORE TO EXPLORE

RESOURCES

Glossary

aquarium (uh-KWAIR-ee-uhm) A building people can visit to look at sea life.

domestic (doh-MES-tik) Tame or living near or with humans.

invertebrate (in-VUR-tuh-breyt) A type of animal that does not have a backbone.

memory (MEM-uh-ree) The power to remember what has been learned.

pattern (PAT-ern) The regular and repeated way in which something happens or is done.

predator (PRED-uh-tuhr) An animal that eats other animals.

primate (PRI-mayt) Any member of the group of animals that includes humans, apes, and monkeys.

Read More

Austen, Lily. *Smartest Animals.* Minneapolis: Jump!, Inc., 2025.

Riggs, Kate. *Chimpanzees.* Mankato, Minn.: Creative Education/Creative Paperbacks, 2022.

Index

TOP RANK is published by Black Rabbit Books, P.O. Box 227, Mankato, MN, 56002.

• Top Rank is an imprint of Black Rabbit Books. • Edited by Alissa Thielges • Designed by Danny Nanos • Photographs © Alamy Stock Photo/Howard Chew, 5, 23; Dreamstime/Altitudevs, 9, 21, Jolanta Wojcicka, 4, 21, Musat Christian, cover, Sashasoloshenko91, cover; Getty Images/Redders48, 11, Westend61, 23; Shutterstock/682A IA, 17, Amelia Martin, 20, 23, andregric, 13, 21, Barbara_C, 18, Bart Groundhog, 15, battybattrick, 23, Christian Musat, 10, 21, Henner Damke, 2, 5, LiskaM, 14, lukaszemanphoto, 6–7, RHIMAGE, 17, 23, Rudmer Zwerver, 19, 21, Smeerjewegproducties, 8, 23, talseN, 16, 21, U__Photo, 12 • Printed in China

Library of Congress Cataloging-in-Publication Data: Names: Mattern, Joanne, 1963– author. | Title: Super-smart animals / by Joanne Mattern. Description: Mankato, MN: Top Rank is an imprint of Black Rabbit Books, [2025] | Series: Super-incredible animals | Includes bibliographical references and index. | Audience: Ages 8–11 | Audience: Grades 2–3 | Identifiers: LCCN 2024017573 | ISBN 9781645823827 (library binding) | ISBN 9781645824046 (paperback) | ISBN 9781645824268 (ebook) | Subjects: LCSH: Animal intelligence—Juvenile literature. | Classification: LCC QL785 .M44 2025 (print) | LCC QL785 (ebook) | DDC 591.5/13—dc23/eng/20240424 | LC record available at https://lccn.loc.gov/2024017573 | LC ebook record available at https://lccn.loc.gov/2024017574